CRYSTALS
What They Are and How to Grow Them

By Marlene M. Robinson

RUNNING PRESS
Philadelphia ♦ Pennsylvania

Copyright © 1988, 1993 by Running Press.
Printed in the United States of America.
All rights reserved under the Pan-American and International
Copyright Conventions.

This book may not be reproduced in whole or in part in any form
or by any means, electronic or mechanical, including photo-
copying, recording, or by any information storage and retrieval
system now known or hereafter invented, without written
permission from the publisher.

Canadian representatives: General Publishing Co., Ltd.,
30 Lesmill Road, Don Mills, Ontario M3B 2T6.
International representatives: Worldwide Media Services, Inc.,
30 Montgomery Street, Jersey City, New Jersey 07302.

9 8 7 6 5 4 3 2 1
Digit on the right indicates the number of this printing.

Library of Congress Cataloging-in-Publication Number 92–50800
ISBN 1-56138-239-6

Package and book cover design by Toby Schmidt
Package and book cover illustrations by Lili Schwartz
Book interior illustration by Bruce Lohr

Package cover photographs courtesy of Department of Library
Services, American Museum of Natural History: (front cover,
clockwise from top:) Neg. nos. K11240, K13972, K13060; (spine:)
2372; (inside flap, clockwise from left:) K14465, 2768, K1306;
(back cover:) 2082.

Book cover photographs courtesy of Department of Library
Services, American Museum of Natural History: (center:) Neg. no.
2768, (clockwise from upper left corner:) K13972, K14465.

Photographs courtesy of Department of Library Services,
American Museum of Natural History: p. 8, Neg. no. 319025
(photo by Thane L. Bierwert); p. 10, Neg. no. 337205 (photo by
AMNH Studio); p. 12, Neg. no. 121940 (photo by Thane L.
Bierwert); p.13, Neg. no. 41954; p. 13, Neg. no. 2A12937 (photo by
R.P. Sheridan); p. 14, Neg. no. 120045 (photo by Thane L.
Bierwert); p. 17, Neg. K5302; p. 18, Neg. no. 247724 (photo by
Julius Kirschner); p. 25, Neg. no. 35919; p. 28, Neg. no. 314040
(photo by Julius Kirschner); p. 29, Neg. no. 2A2394 (photo by Lee
Boltin); p. 31, Neg. no. 327471; p. 37, Neg. no. 336943 (photo by
Singer); p. 38, Neg. no. 315590 (photo by Julius Kirschner); p. 38,
Neg. no. 117703 (photo by Julius Kirschner); p. 38, Neg. no.
124806, (photo by Rota); p. 42, Neg. no. 2A8857; p. 50, Neg. no.
328170 (photo by Rota); p. 54, Neg. no. 41971; p. 56, Neg. no.
125737 (photo by Rota); p. 57, Neg. no. 2A12934 (photo by R.P.
Sheridan); p. 62, Neg. no. 3133 (photo by Weber); p. 63, Neg. no.
3121; p. 64, Neg. no. 125733 (photo by Rota); p. 64, Neg. no.
2A9041; p. 65, Neg. no. 335676 (photo by A. Singer); p. 66, Neg.
no. 297287 (photo by Dwight Bentel); p. 67, Neg. no. 328796.

Photographs courtesy of AT&T Archives: pp. 27, 32, 80, 82; The
Bettmann Archive, pp. 48, 55, 59, 60, 61; The Granger Collection,
p. 24; Hawaii Visitors Bureau, p. 16; Kristal Corp.: pp. 1, 23, 73;
National Aeronautics and Space Administration: pp. 47, 52, 84;
Smithsonian Institution photo no. 788853A, pp. 36, 40.

Typography by Commcor Communications Corporation,
Philadelphia, Pennsylvania.

Running Press Book Publishers
125 South Twenty-second Street
Philadelphia, Pennsylvania 19103

Contents

Growing Your Own Crystals 5

1. Hunting for Crystals 8
   Try It Out: Building Super Salt Crystals

2. Stone Spears and Silicon Chips 24
   Try It Out: Making Rainbows
   with Crystals

3. Million-Dollar Crystals 36
   Try It Out: Diamonds and Graphite
   from Gumdrop Models

4. On the Shoulders of Giants 47
   Try It Out: Mohs's Scale of Hardness

5. How Nature Makes Crystals 55
   Try It Out: Finding Crystals in Rocks

6. Crystal Family Resemblances 68
   Try It Out: Classic Crystal Family
   Shapes

7. A New Age for Science 80

   Glossary 86

   Index 87

> Do not swallow chemicals, solutions, or crystals. Avoid contact with eyes and mouth. Keep out of reach of small children.
> Read the following instructions before use.

Growing Your Own Crystals

With the packet of monoammonium phosphate included in this kit, you can grow two small clusters of sparkling yellow, quartz-like crystals.

Like many kinds of crystals that grow within the earth, these crystals form by a process of cooling.

For this activity, you'll need the packet of monoammonium phosphate included in this kit; a small plastic bucket; boiling water; a container 4" wide and 7" or 8" deep (such as a plastic cola bottle with the top cut off) to hold the crystals; an old saucepan; a clean twig or small paintbrush with a long handle to stir the solution; two small, clean rocks (stones about 4" in diameter and 1½" tall with rough surfaces are best, but don't use cement); and a pair of rubber gloves.

1. Empty the packet of monoammonium phosphate into the bucket.

2. Add 1¼ cups (10 ounces) boiling water, measuring carefully. (Crystals won't grow if you add too much water.) Stir, using the twig or the handle of a paintbrush, until the monoammonium

phosphate dissolves into a yellow solution.

3. Place one of the rocks into the crystal container and pour in the yellow solution. The solution should cover the top of the rock by at least ½".

Place the container where it will be undisturbed, out of the reach of pets and small children, and where the room temperature is fairly constant. Within two or three hours, crystals will begin to form on the rock and on the bottom of the container.

4. Let the crystals grow for about one week.

(If no crystals grow, too much water was used. Pour the solution into the saucepan, boil a few minutes, and then repeat step 3 above to grow beautiful crystals. Be sure to wash the saucepan thoroughly after use.)

5. To remove the crystals from the container, wear rubber gloves to protect your hands from the yellow dye. Remove the rock from the solution, put it on a sheet of newspaper or paper towel, and let it dry for a few days, away from water and excessive heat.

6. You can use the remaining solution in the crystal container to grow a second cluster of crystals.

Pour the solution into the saucepan,

together with any small crystals that grew on the bottom or walls of the container. Heat until dissolved.

Rinse out the crystal container and repeat steps 3 through 5 to grow another marvelous crystal.

Wash the saucepan thoroughly after each use.

◆ ONE ◆
HUNTING FOR CRYSTALS

What is a crystal? A dazzling diamond. A grain of sand. A snowflake. A fleck of silver, or a speck of salt.

All these, and millions more, are crystals. You live in a world filled with crystals. You write with crystals, tell time with crystals, eat and drink crystals, and sometimes bathe with crystals.

Quartz crystals

Most of the crystals around you are so small that you never notice them. Some are so common, you'd never think they could be so fantastic. Others are so rare that fortunes and even lives have been gambled just to win possession of them.

What is a Crystal?

There are many scientific ways to identify and define crystals, but your senses of sight and touch are the only tools you need to begin exploring the wonders of crystals.

First, of course, you'll need a crystal. Use one you've grown from the materials provided with this book, or use a crystal you have found or purchased from a rock or hobby shop. If you don't have a crystal handy, study the picture of the crystal at the beginning of this chapter and use your imagination to experience how it would feel in your hand.

Now pick up your crystal, read the questions below, close your eyes, and concentrate.

- When you squeeze the crystal, does it feel warm or cool?
- Does the shape change?
- Are the surfaces of the crystal smooth or rough? Are they all the same shape? How many are there?
- Are the edges of the crystal rounded or sharp? Do they come together in points? If so, how many?

Open your eyes. Now is a good time to start a crystal notebook to record your observations. This notebook can be a special place to keep all the amazing things you'll discover about crystals throughout this book.

One of the first things you may notice when you pick up a crystal is the temperature of the stone. Certain minerals can be iden-

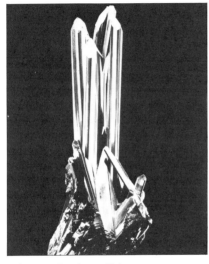

Stibnite crystals with a metallic luster

tified this way. For example, real jade is always cool. A pure quartz crystal feels cold and glassy.

Another key characteristic is the crystal's hardness. Most crystals are hard, but some, such as graphite, are so soft you can scratch them with your fingernail.

Does your crystal have flat sides, or is it textured? Sometimes you can actually feel characteristics that reveal a crystal's atomic structure. Faint grooves on crystal surfaces are one sign of its inner structure; another is in the number of surfaces, or faces, a crystal has.

By now you probably have discovered some of the basic characteristics of a crystal: straight edges, clean angles, clearly defined surfaces, a particular hardness, and a predictable shape.

So what is a crystal? A chemist who answers that question may focus on the minerals found in a crystal. A geologist could define a crystal by the geological forces that formed it. A research scientist might discuss the molecular arrangement inside the crystal.

All of these experts, however, would agree on this definition: A *crystal—whether a metal, gemstone, or mineral—is solid matter showing a symmetrical pattern of faces and angles caused by an orderly, repetitive arrangement of atoms.*

Armed with these simple facts about crystals and a magnifying glass, you can find hundreds of crystals under your nose—or your toes! Serious crystal hunters should see the suggestions on pages 18 and 19 for the best ways to look at these minuscule wonders through a magnifying glass, but for now let's consider where you'll find the crystals that surround you every day.

Indoors

The kitchen is chock full of crystal shapes. Compare, for example, table salt and sugar. Did you realize you were eating cubes on your french fries? Or triangles on your cereal? Sugar crystals take the shape of rectangles, triangles, or tiny multi-sided columns.

Similar substances can have very different crystal shapes; consider rock salt, kosher salt, and sea salt, or brown sugar, powdered sugar, and fructose. When you find differences among the shapes of these crystals, it means there are chemical differences in the crystal makeup. For example, sea salt is a combination of several kinds of salt.

Compare sugar with sugar substitutes; saccharin contains several ingredients showing a mix of rectangular and long slabs with triangular cross sections. Containers of maple syrup, honey, molasses, and soy sauce can develop crystals if they sit long enough.

Alum, used in pickling, usually looks like two pyramids glued back to back. Baking soda is crystalline, which means its crystal structure can only be seen with a microscope.

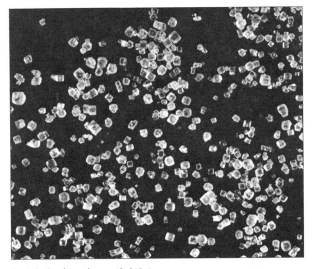

Crystals of cooking salt, magnified 12 times

But if you hold it in just the right light, you may see tiny crystal faces.

Dry mixes such as lemonade, powdered soft drinks, camping foods, or instant iced tea contain several types of crystals, including those of preservatives. Freeze-dried coffee and soup may hold crystal shapes.

Freezer frost forms quickly in frosty crystalline needles and pillars.

In the bathroom, you might find bubble bath powders with crystalline substances and ingredients with clearly defined crystal shapes. Look in the medicine chest for epsom salts with long, pointed, branching crystals, and boric acid that shows flat, shiny, triangular crystals.

Metal objects throughout your

home also could reveal crystal shapes. Inspect doorknobs, silverware, pots, vases, lamps, picture frames, or keys with a magnifying glass. If metals are highly polished, they may reveal crystal structures. Clear differences emerge among iron, brass, steel, silver, gold, and aluminum crystals.

Crystal goblets are indeed made of quartz crystal, but because the quartz has been melted down, you do not see crystals when you inspect a glass.

If you carefully open the back of an old watch, you might find crystal bearings of ruby or quartz. Jewelry, especially if it contains gemstones, is another source for a cache of crystals.

Copper is occasionally found in crystal formations (top); Gem-quality smoky quartz from Brazil

Of all the building materials, granite has the most varied crystals.

You might say that some crystals around the house have the "write" stuff. Pencil lead, though soft, is composed of microscopic crystals of graphite. LED displays are made of a thin layer of liquid crystals.

In a basement, workshop, or garage, there may be thousands of tiny diamonds in the diamond coating used to increase sharpness and extend wear on drill bits and saws. Sandpaper contains various sizes of crystal shapes. Plant food also contains crystals, usually blue and clear ones.

Outdoors

If you take your crystal expedition outdoors, the variety of crystals to discover really expands.

Building materials contain all kinds of possibilities for crystal hunters. In the East, granite and brownstone are popular building materials, and a freshly chipped spot is the best place to see crystal shapes. Flagstone is used in gardens all over the country, and as a building material in the West, along with local sandstones.

Of these, granite has more varied and visible crystals, with quartz being its major ingredient. Concrete blocks contain sand, and a few crystals of it may be visible.

Driveways, gardens, water filter tanks, and sandboxes may hold construction sand. This coarse, yellow-brown sand won't look much like crystals; it is gathered from stream beds where the tumbling action of water has worn down the edges. The brownish particles in it usually are hematite, an iron-rich mineral with a globular crystal form.

White playground sand has smaller, battered particles with a variety of quartz colors: white, pink, purple, smoky, yellow, green, or clear. A determined crystal hunter may find pink or white feldspar and black or gold mica. If tiny black crystals cling to a magnet, they are magnetite or ilmenite.

The sand under your toes tells a centuries-old story of crystal formation. All sand was once something else—look to the rocks, cliffs, or mountains around you for clues.

If you're on the beach in Florida, there's no natural quartz sand underfoot. Florida beaches are made of ground-up coral and seashells, but most other beaches on both coasts are quartz. In Hawaii, there are both black and green beaches because black obsidian and green olivine rocks spewed out of volcanoes and, over the centuries, were ground to sand.

Black sand at Kalapana, on the island of Hawaii, was formed from volcanic rock.

In desert sand, the grains have been blown about so much by the wind that they resemble tiny crystal balls!

The kinds of crystals found in local rocks depend upon what happened in your area millions of years ago. The forces that thrust up mountains and wore them down again determine the types of crystals to be found. Consult a rock and mineral guide for the best places in your area to search for crystals.

Crystal hunters who live in the snowbelt have an extra source of crystals in the form of snowflakes. If you can catch some snowflakes and look at them through a magnifying glass (for best results, use a black velvet background), you would see that each one has six sides. An estimated one million snowflakes fall in a 2-foot area during a 10-inch snowfall.

One common belief about snowflakes is

that no two are alike. That bit of folklore has been challenged by a researcher for the National Center for Atmospheric Research in Colorado who photographed two remarkably similar snow crystals. Still, it is rare to find the same pattern among the millions of flakes that fall in every storm.

Snowflakes are crystals of water that condense on dots of microscopic dust. If our atmosphere were dust-free, we would have no snowflakes. The size and intricacy of each pattern depend on the range of temperature and moisture the flake passes through on its way to earth.

Don't forget to compare these winter wonders with freezer or window frost.

The Crystal Hunter's Field Guide

Your journey into the world of crystals works best if you have a magnifying lens of at least 5-power. If you have a microscope, that's even better.

Before launching your crystal expedition,

Snowflakes are the world's most beautiful ice crystals, but scientists are no longer certain that each one is unique.

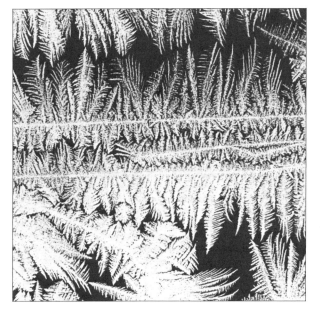

remember that when using a magnifying lens you must bring specimens close to your face. Do *not inspect any strong or poisonous substances,* such as drain or toilet bowl cleaners. If in doubt, ask an adult for advice.

Explore every room indoors, and then head outdoors to hunt for more crystals. Here are some tips to successful crystal gazing:

- Inspect each sample with a magnifying lens, turning pieces with a toothpick to see shapes clearly.
- Make sure you have a strong source of light.

Compare this window frost with the crystalline needles and pillars of frost in your home freezer.

- Look for straight edges, flat faces, and repeated shapes. Even if you think you have found a perfect crystal, look at several samples to make sure you have a typical specimen.
- Transfer a powder or loose material to a dark background to make it more visible. A small tray, sheet of black construction paper, a frying pan, or a piece of black cloth (especially velvet) work beautifully. (Snowflakes or freezer frost also last longer if you use a black cloth background.)

To organize your search, enter these four headings in your crystal notebook: Crystal, Crystalline, Organic, and Unknown.

If you can see crystal shapes and angles in repeated, consistent patterns, the specimen belongs in the crystal category. If crystal shapes are not visible, put the specimen into the crystalline category—unless it once was a living thing.

Substances that were once living are organic. Organic materials contain chemicals in crystal form, but cells are their basic building blocks and protein their building material.

Plastics are technically organic, even though they are manufactured, because they are made from petroleum. Ancient plant bogs, and the prehistoric creatures that lived and died in them, distilled over millions of years into oil and gas. When scientists discovered how to spin long molecules of oil, called polymers, today's plastics industry was born.

The Unknown category is for other things you can't easily classify and will have to research.

Try it Out

Building Super Salt Crystals
Making supersized salt crystals at home is easy. You can measure the ingredients precisely, or just dump and stir. Either method usually works.

Before you start, stop to think about safety. You'll need to use the kitchen stove. Don't set anything flammable on the burners, make sure you have a potholder handy, and be extra-careful when handling hot liquids. Don't use a microwave oven for this experiment because you'll need to stir the solution while it is heating, and you'll want to watch.

To grow a salt crystal you'll need:

2 ½ cups table salt
1 cup water
12″ of thread (cotton is best; don't use nylon)
pencil
wide-mouthed glass jar, such as a peanut butter jar
tape
paper towel

Combine the salt and water in a pan and stir. See how much salt you can dissolve by stirring. When the water holds all the salt it can at room temperature, it's called a saturated solution.

Now heat the solution, but don't let it boil. Watch the remaining grains of salt. They will be absorbed into the solution as the temperature rises, but see if this happens more quickly when you stir.

When all the salt has dissolved into the saturated solution, add more. Measure the amounts you add—tablespoon by tablespoon—to see how much more salt can be dissolved into hot water than into cold water. Now you have a *super*-saturated solution (SSS for short).

Remove the pan from the stove to cool for a few minutes. Tie one end of the thread around the middle of the pencil, tape the other end of the thread to the bottom of the jar, and lay the pencil across the top of the jar as shown (figure 1).

Carefully pour the cooled SSS into the jar. Tape a sheet of paper towel over the top to keep out dirt. Put the jar where you can see it without moving it. An ideal spot is in a cupboard where the temperature remains the same. If you put your jar in a sunny window, the heat from the sun will speed up evaporation and only tiny crystals will form.

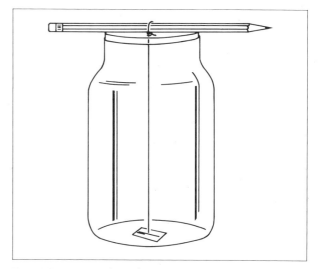

Figure 1: *Growing super salt crystals in a jar*

During the next few weeks, check your jar every day for crystals. If nothing happens, drop a grain or two of salt into the jar to act as "seeds" to get your crystals going.

If you want to make larger crystals, keep one perfect crystal on the thread as the seed. Pull the string out of the solution and carefully remove the crystals you don't want to preserve by tapping with a small hammer. Don't tap near your perfect crystal or you might accidentally knock it off the string.

Pour the solution back into a saucepan, add any crystals you can scrape from the jar, plus the ones you knocked from the string, and heat the solution until all the salt has dissolved. (The extra crystals will dissolve more quickly if you crush them.)

Shorten the thread so that your perfect crystal hangs just above the bottom of the jar. Cool the solution to room temperature and pour it back into the jar. Make sure the solution is cool; if it's too hot, your seed crystal will melt. Cover the jar with a paper towel and watch your crystal grow during the next few weeks.

Crystals of salt will form on the sides and bottom of the jar, but some will cling to the thread where it's easier for you to inspect them. Notice how they resemble the tiny cubes of table salt. How many faces do you see? Six. What kinds of angles? They are all 90 degrees.

A salt crystal grows at the rate of at least 100 layers of atoms every second. This pattern continues until something gets in the way and the growth changes direction, or until growing conditions are no longer favorable and crystal formation stops.

When crystals grow together and interrupt each other's development, their cube shapes are incomplete. This is why scientists want to grow crystals in space, where they won't bump into each other and where gravity won't pull them to the bottom of a container.

The fact that your large crystals of salt have the same shape as tiny crystals of table salt shows that your crystal has a blueprint for growth. The atoms of a crystal fit together like pieces in a puzzle, and their orderly arrangement creates a beautiful shape.

This recipe for growing salt crystals can be used to enlarge crystals of other substances you found on your crystal hunt. Epsom salts, boric acid, alum, sugar, monosodium glutamate, or plant food are easy and safe to grow. If you want to try others, check with an adult first to avoid harmful chemicals.

By the way, sugar crystals are the old-fashioned "rock candy" treats that your great-grandparents used to buy. If you are careful to keep other chemicals out of the way when making rock candy, you can eat your experiment!

You can conduct more investigations with all of these supersized crystals, so don't throw them away.

An alum crystal is one of many crystals you can grow at home using common ingredients.

◆ TWO ◆
STONE SPEARS AND SILICON CHIPS

What do stone tools, microwave ovens, and crystal balls have in common? What helps run computers, VCRs, automobile systems, and some watches? Quartz—one of the most common minerals on earth.

The chemical name for quartz is silicon dioxide. You may be most familiar with this plentiful crystal in the form

This woodcut was carved in 1530, about 80 years before Galileo proved that planets revolve around the Sun.

of sand. Sand, and its silicate relatives, make up more than 90 percent of the earth's crust.

Quartz has played a role in every major technological advance for nearly two million years. Today, quartz has more uses than any other solid mineral and it has so revolutionized our lives that the second half of the 20th century could well be called the Age of Quartz.

Stone Age Inventions

Sometime between two million years ago and one-and-one-half million years ago, one of our Stone Age ancestors pounded on a fist-sized river rock—a chunk of dark-colored crystal quartz we know as flint—and produced the first tool.

Next, Stone Age people made axes, scrapers, spears, and finally arrowheads. To make these tools

---

Life in the Stone Age was revolutionized by flint tools like this one.

they used chert and jasper—two other types of quartz—as well as other minerals, but flint was the most common.

Flint tools revolutionized the lives of Stone Age people. Hunters could kill larger animals, skin their prey, and make crude sleds for transporting food and skins.

Seed gatherers used flint blades to strip reeds and to harvest grasses that could be used to weave baskets and make clothing and shelter. With flint blades, wood could be cut and bound into bundles for carrying. Flint tools made tasks faster and easier; they made a new kind of life possible.

Throughout ancient Africa, Europe, Asia, and America, there were plenty of chunks of flint to be found along rocky stream beds and in sandstone cliffs.

Flint was the best material for making stone tools because it was hard enough to hold a cutting edge, yet easily shaped. Small pieces of flint could be chipped off in predictable patterns that let Stone Age hunters control the thickness, length, and shape of their tools.

The Silicon Revolution

It seems incredible that the mineral most valuable to a Stone Age hunter could be the pivot of 20th-century technology. Like Stone Age toolmakers, today's scientists count on the abundance, strength, and durability of quartz.

Scientists also know that silicon—a major component of quartz—is an excellent base for the electrical circuits, diodes, transistors, and resistors etched into silicon chips.

The silicon chip ushered in the second industrial revolution in 1955. Look around you: VCRs, microwave ovens, calculators, TVs, computers, automobile systems, industrial robots—even artificial arms and legs—function because of a fleck of quartz crystal.

In the 1950s, the silicon chip was a quarter-sized wafer; in the 1960s, it shrunk to the size of a postage stamp; in the 1970s, it was small enough to fit through the eye of a needle; and in the 1980s, it dwindled to the size of a grain of salt.

As the chip got smaller, scientists increased the number and speed of its circuits — the pathways and gates that carry or detour electrons. Silicon chips in some home computers now have as many as 16 million circuits. Each is thinner than a human hair and the circuits are etched with a laser.

This dime-sized silicon chip has as much processing power as a minicomputer.

An Asian artist carved both this elephant and the polished ball from rock crystal.

Crystal-Gazing

The ancient Egyptians, who lived on the edge of the Sahara Desert, must have seen that intense heat from fire or lightning could fuse sand into a glassy glob. They developed a process for melting and molding sand, and the beautiful bottles they formed were the beginning of glassmaking. Fine crystal goblets, as well as windowpanes, are still made from quartz sand.

Quartz glass is also used to make high-quality fibers for fiber optics, which have improved telephone systems and photocommunications. Surgeons can now perform delicate operations with the aid of a tiny camera attached to the tip of an optic fiber to show the way.

Sparkling crystals that catch and scatter the light—especially quartz—have been collected as gemstones and charms for thousands of years. For centuries, fortune-tellers and soothsayers have claimed that by gazing into crystal balls shaped from clear rock quartz, they can foresee the future and reveal hidden secrets.

Some people must have noticed how certain crystals and gemstones magnified objects. It is said that the Roman Emperor Nero improved his vision by holding a large, curved jewel in front of one eye.

But a distrust of inventions delayed the eye-opening ways that quartz glass could improve sight. Alchemists of the Middle Ages believed more in the sense of touch than of sight, and it's no wonder, for they lived in dark castles lit by candles and torches.

Unless sunlight was streaming into their workshops, the alchemists could not see very well, so they had to count on a good sense of touch. With their natural suspicion of inventions, they mistrusted anything observed through a curved lens.

Sometime between 1280 and 1285, quartz finally came to the rescue as glass spectacles with curved lenses were invented.

Some crystals, such as gypsum, can be used as magnifiers.

They were held up to the eyes, but later were fitted to the nose. Lenses are so named because people thought they looked like the small, flat beans called lentils. For more than 300 years, eyeglasses were called "glass lentils."

Think how these crude spectacles must have changed people's lives! Not only did they allow men and women to go about their work more easily, but they must have provided a new sense of independence to the weak-sighted.

Roger Bacon (1214–1292), the most celebrated European scientist of the Middle Ages, was a monk who spent his life trying "to work out the natures and properties of things." He is believed to be one of the first to grind quartz glass into a lens that would magnify objects.

Through a series of experiments (like those you'll find on page 34), Bacon discovered why some lenses cause blurry or double vision, and why others reflect color into the eyes.

Unfortunately, rather than becoming famous for his discoveries, Bacon was considered a wizard of black magic and imprisoned.

Nearly 300 years after Bacon's discoveries, society was finally ready for lenses. By then, the idea of scientific investigation was accepted, and the optical properties of quartz opened the way for a new era of invention and technology.

In the late 1500s, people began using the first telescopes. They were the size of a thumb, with a concave lens at one end. They were called "flea glasses," and you can guess what their owners used them to look for.

About 1608, the Italian scientist and

philosopher Galileo built a better telescope and with it proved that the earth and other planets revolve around the sun.

By the end of the 17th century, experimental science had come into its own, led by Isaac Newton. And what a difference 500 years made! Roger Bacon went to prison for his experiments, but Newton was glorified as a hero and knighted for his scientific discoveries.

One of Newton's first great accomplishments was to continue the work of Bacon on the nature of light. Newton used crystal quartz glass, prisms, and lenses to prove that white light holds a spectrum of rainbow hues.

Galileo used this telescope to prove that the Earth and other planets revolve around the sun.

Hundreds of integrated circuits, like those in the background, fit on a slice of semiconductor made from silicon.

'Dancing' Crystals

If you hold a crystal in your hand, it does not seem to move. Inside that elegant shape, however, the crystal's atoms are doing a dance, just as all atoms everywhere are in constant motion.

The movement of quartz atoms is different from that of other minerals. Quartz atoms vibrate so evenly, so regularly, and so predictably that scientists use this movement as a measuring tool. In fact, the quartz measure is the world's most accurate scale.

By pairing a super-thin slice of quartz with a material to be measured, scientists note how the quartz vibrations change. Quartz has been used to measure everything from specks of dust to atomic explosions.

If you squeeze a quartz crystal, you create an electrical charge. This discovery led to the development of the first radios, or crystal sets, in the early 1900s. Sounds could be heard when the frequency of radio waves matched the vibrations of the quartz crystal as they passed through it. In today's telephones, thin quartz crystals tuned to different frequencies keep simultaneous calls separate.

Quartz opened up the field of electronics when silicon transistors were developed. They replaced radio vacuum tubes, which were large and required a great deal of energy. Silicon chips shrunk radios from tabletop models to pocket size.

Even versatile, dependable quartz has its limitations. When overheated, it loses its ability to conduct electricity. This is why every electronic device using silicon chips, such as a computer, has a cooling system to suck cool air across the chips to keep them operating at peak efficiency.

High-tech industries, aware of silicon's heat limitation, have begun to find replacements. Bismuth, lithium, yttrium, and gallium are overtaking quartz as semiconductors, and semiconductors themselves are losing ground to superconductors— elements other than silicon with clearer, cleaner alignments of molecules creating barrier-free paths for conducting electricity.

Does this mean we have reached the end of the Age of Quartz? The answer, of course, is no. Scientists are still discovering ways to use quartz in medicine, industry, and research. After more than a million years, we have not yet exhausted the possibilities in a grain of sand.

Try it Out
Making Rainbows with Crystals

Imagine that you are the good monk Roger Bacon, living in the Middle Ages. You are curious about the nature of light, so you experiment to see how light travels through objects.

To try this experiment you will need:

light source (a flashlight fitted with a paper cone, such as figure 2); a movie or slide projector fitted with a slotted mask; or any other source of light that can be focused in a beam)

white background (a white wall or a sheet of white paper or cardboard propped against a wall)

objects for experimentation (choose several)

glass of water
clear plastic bottle of seltzer or soda
other clear containers of liquid (syrup, pickle juice, or milk)
marbles
exposed film
a pair of eyeglasses
magnifying glass
light bulb
several small mirrors
crystals you have grown
crystals of quartz or other minerals

Do what Roger Bacon did 700 years ago—try things out, experiment, have some fun. Darken the room and shine the light through the objects you picked and onto the white background.

As the light beams through different objects, find out where it goes. Describe how it looks before and after. Does it always strike the white background? Does any object change the appearance of the light that

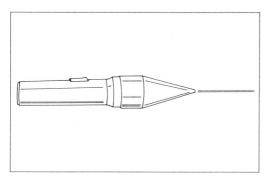

Figure 2: A paper cone will help focus light for experiments with crystals.

passes through it? Does light change the appearance of an object?

Write your observations in your crystal notebook. Make a separate column for objects made of quartz glass. Don't forget to test mirrors.

Like Roger Bacon, you'll probably discover that:

- When a beam of white light travels through some objects it becomes a rainbow of colors. Some of these objects are curved quartz glass, prisms, or quartz crystals. You may have found others.
- When light passes through some objects, it bends.
- When light hits a mirror, it does not go through, but bounces off at an angle.

◆ THREE ◆
MILLION-DOLLAR CRYSTALS

Blazing diamonds, fiery opals, gleaming gold and silver, crimson rubies, sparkling sapphires: all breathtaking gems and metals owe their existence to the formation of crystals. Throughout history, kingdoms have been won and lost, undying love has been declared, and great leaders have been honored and rewarded with precious crystals.

The 45.5-carat Hope diamond

For thousands of years, gemstones have been an expression of wealth. In the United States, diamonds are the most highly prized crystals. But other countries and cultures value a variety of gemstones.

In India, rubies have long been the most sought-after stone, but since so many precious gems are found there, the competition includes sapphires, emeralds, and diamonds.

When India was divided into many kingdoms ruled by rajahs and princes, one hefty ruler demanded an annual tribute of his weight in precious stones and metals. An enormous scale was built in a public square. As he sat on one side, his subjects piled gems and gold on the other until their weight swung him off the ground. China has had a love affair with jade for more than 5,000 years. Beautiful, highly intricate sculpture, jewelry, vases, and ritual implements have been carved from this dense, crystalline stone, which comes in lustrous shades of green, pink, yellow, and white. Jade conveys feelings of serenity and durability,

Emerald crystals match those of diamonds for beauty, and are even more rare.

Kannon, the Goddess of Mercy to Japanese Buddhists, carved from pale green jadeite (left); a dagger with a jade handle from India (top right); a Chinese white jade cup with handles carved in the shape of dragons.

which may explain why one ancient Chinese princess was buried in a suit of armor fashioned with jade.

Green malachite, blue azurite, deep blue lapis lazuli, and coral-orange carnelian were the precious stones of ancient Ur and Egypt. More than 5,000 years ago these gemstones were mounted in elegant gold settings as tiaras, neck collars, and bracelets for the living and the dead.

The Mountain of Light: Koh-i-noor

The cause of more bloodshed and intrigue than any other gem probably was the Persian Koh-i-noor, or "Mountain of Light" diamond. For centuries, it shifted through Asia and the Middle East, passing through the hands of Indian rajahs, Mogul emperors, and freewheeling adventurers.

In 1739, a Persian shah followed his country's custom of trading his turban with that of the ruler he had just defeated. Imagine the victorious shah's surprise when he discovered the 186-carat Koh-i-noor diamond hidden in the folds of cloth!

In the 19th century, the Koh-i-noor passed into the hands of an Afghan prince, who believed that it brought him luck. He so dearly wanted to keep the dazzling diamond that he endured torture and was blinded before surrendering it.

When Great Britain annexed the Punjab province of India, the Koh-i-noor was presented to England's Queen Victoria. To improve its brilliance she had it cut down to 109 carats, but her craftsmen botched the job. No matter, Koh-i-noor still became the centerpiece of the crown of the Queen Mother of England.

The Hope with a Curse?

The most famous diamond in the United States is the Hope diamond, a deep blue, 45.52-carat stone that reflects a million fiery lights. Legend has it that the Hope was cut from the famous 112-carat Blue Tavernier diamond purchased by a French merchant, Tavernier, in 1642 from the Kollur mine in India. Others say that it was stolen from the statue of an Indian god by a man who was later devoured by tigers.

It is known that in 1668, Tavernier sold his gem to King Louis XIV of France, who proclaimed it "The Blue Diamond of the Crown." French lapidaries recut it to bring out its brilliance, reducing the stone to 67.5 carats. It was one of the jewels stolen during the French Revolution and never recovered— unless, of course, it survives as the Hope diamond.

A diamond of similar color, but only 45.5

The Hope diamond: a fabulous crystal with a notorious past.

carats, was bought in London in 1830 by Henry Thomas Hope, an American banker, and this stone was assumed to be the "lost" French diamond.

In the United States, it was called the Hope diamond, but its reputation mocked its name. Hope's son lost his fortune after inheriting the stone. In the early 1900s, it was sold several times before being purchased by Mrs. Edward A. McLean.

The McLean family suffered considerable misery after the purchase, and the Hope later was sold to New York diamond merchant Harry Winston. After many tries, he found no buyers for the jewel, which he donated to the Smithsonian Institution in Washington, D.C., in 1958. At that time, it was valued at $1 million and it is still the most popular exhibit in the 13-building museum.

Diamond in the Rough

In 1905, a glint of yellow caught the eye of the manager of the Premier diamond mine in South Africa. Sure of a prank, he dug into the mine wall with his pocketknife, and extracted a 1 ⅓-pound glittering stone!

The rough diamond was nearly 4 inches by 2 ½ inches and a crystal of exceptionally fine quality weighing 3,106 carats— the largest rough diamond ever found. Named for the owner of the mine, the Cullinan Diamond was cut into two large crystals and many smaller gems.

The pear-shaped Cullinan I, or "Star of India," weighs 530.20 carats and has 74 facets. It was presented to England's King Edward VII on his 66th birthday in 1907, and later was set into the Royal Scepter. Cullinan II weighs 312.40 carats, has 66 facets, and was placed in the Imperial State Crown. Both are

Glass model of the Cullinan diamond before it was cut into two large crystals for the British Crown Jewels — including the Star of India, which was set into the Royal Scepter.

among the British Crown Jewels that are kept locked in the Tower of London.

Try it Out

Diamonds and Graphite from Gumdrop Models

The value of an uncut diamond crystal depends upon its hardness, how it reflects light, and its "fire"—the way it breaks up light into prismatic colors.

You would think a substance so valuable and durable would be made of special elements, but the stuff that makes diamonds is about as rare as the lead in your pencil. In fact, it is composed of exactly the same element: carbon.

How can one element be as hard as a diamond, or as soft as the graphite in pencil lead? The angles between atoms and the length of the bonds between the layers of atoms make the difference—the difference between a glittering diamond and a pencil point of dull, gray lead.

You can see this for yourself by building gumdrop models of the atomic structures for diamond and graphite. Let's start with a diamond, which shows perfect atomic symmetry.

To make a model of a diamond you will need:

a protractor
30 gumdrops
a box of toothpicks
your crystal notebook

Lay the protractor along a line in the middle of a notebook page and mark an angle of 110 degrees from the center hole of the protractor. Mark the base line extending from this center hole. Measure 2 inches along each line and place a gumdrop at these points and in the center of the angle (figure 3). Draw

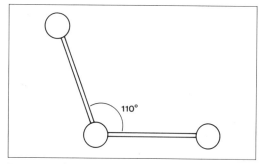

Figure 3: Template to create 110° angles

around the gumdrops. This will be your template for making a gumdrop diamond crystal.

Connect these three gumdrops with toothpicks, breaking off excess wood. Keeping two gumdrops in place, turn the construction so that one points straight up, allowing you to add a fourth gumdrop at the proper

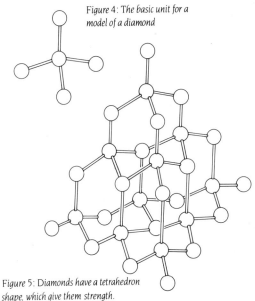

Figure 4: The basic unit for a model of a diamond

Figure 5: Diamonds have a tetrahedron shape, which give them strength.

angle. Stand this tripod up and insert a toothpick and gumdrop into the topmost point (figure 4). This is the basic unit of a diamond— one atom tightly bound to four surrounding atoms.

To make a multifaceted diamond model, connect 30 gumdrops in three layers of unit cells (figure 5). The top layer should contain one unit cell; the second layer should contain three unit cells; and the third layer should contain five unit cells.

When connecting these layers, build all the unit cells first to make sure they join at the proper angle. You will remove six extra gumdrops to make the connections. This model represents half a crystal; if you want to complete it, add layers in reverse order: 5,3,1.

Note how strong your model is! Indeed, the tetrahedron (four-sided) shape is the strongest and most rigid structure known in nature. At least one scientist has proposed using the diamond crystal model to build strong, lightweight platforms for large telescopes, buildings, and space stations.

To make a model of a graphite crystal you will need:

> protractor
> 57 yellow gumdrops
> toothpicks
> your crystal notebook

On a page opposite the diamond crystal, make a template for the graphite crystal. Place the protractor on a line in the middle of the notebook page and mark off an angle of 120 degrees. Along the lines, mark off two inches as before and draw around three gumdrops.

After connecting three atoms, turn the model and connect three more at the proper angle, but all on the same plane so that you end up with a hexahedron (six-sided) ring (figure 6).

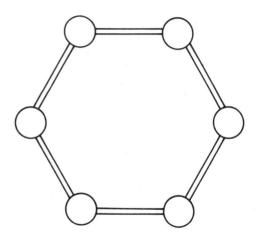

Figure 6: *Hexahedron ring of gumdrops*

Three layers of graphite rings would have five rings to a layer with 19 gumdrops per layer. Since the connections between layers are longer, use the full length of a toothpick (figure 7).

Your model demonstrates that graphite has a weak inner structure; its layers literally slip away from each other, leaving a trail of graphite atoms behind as you write with a pencil.

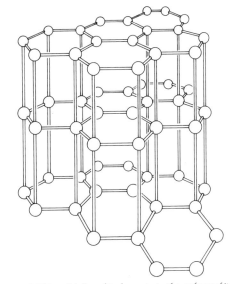

Figure 7: This model of graphite demonstrates the weakness of its atomic structure.

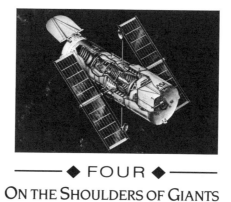

◆ FOUR ◆
ON THE SHOULDERS OF GIANTS

As charms, as jewels, and as workhorses of industry, crystals are fascinating. Who can explain why they are so beautiful? Why they are nearly perfect?

We could not answer these questions today without "standing on the shoulders of giants," as the 17th-century scientist Sir Isaac Newton once acknowledged in a letter

Studying the optical properties of crystals, scientists over the centuries have made discoveries that led to the design for the Hubble Space Telescope.

Medieval alchemists chopped crystals and melted minerals in search of a substance that would transform common metals into gold.

crediting great thinkers before him with discoveries that made his work possible.

Once there was a time when no one knew what the stuff of the universe was. In the Western world, the quest for this knowledge began around 3,000 years ago with the Greeks. In those days, no distinction was made between science and philosophy, so deep thinkers used reason, rather than scientific investigation, to explain the world around them.

By 450 B.C., the Greeks had concluded that the universe was composed of earth, water, fire, and air. Because the great Aristotle promoted it, this belief dominated philosophical science for the next

1,500 years, despite another theory— the atomic theory—that was put forth about the same time.

The atomic theory was developed by a philosopher named Democritus, who concluded that the universe was made up of *atomos* ("indivisible particles"). Democritus realized that these particles were invisible and reasoned that they had different sizes, shapes, and arrangements.

Democritus also theorized that atoms moved about continuously and, in doing so, combined to create objects. If they separated, he concluded, the object died.

Although he could not test his ideas, Democritus produced a theory that holds up amazingly well today. But he did not have the reputation or following of Aristotle, and his atomic theory was largely ignored for 2,000 years.

Mysteries of the Middle Ages

One of the earliest myths about rocks and minerals was that trees were the parents of quartz crystals because quartz may be found embedded in a tree's roots. The ancient Greeks also believed that metals grew and reproduced themselves in the core of the earth.

This became the foundation for the medieval belief that most metals were young and eventually would mature into "noble" metals: lead eventually would turn into silver, and silver into gold.

Medieval alchemists spent their days chopping crystals and melting minerals in search of the secret that could transform matter into the noblest of all substances—gold. They sought the Philosopher's Stone, an imaginary substance that they believed could change common metals into gold.

Medieval scholars believed that geodes were magical omens.

In the meantime, medieval scholars and peasants alike acted on clues they "read" in the stones. Sparkling jewels discovered inside dull brown balls of rock (geodes) were thought to be magical omens, and the flashing fire in an opal was believed to be set by underground spirits.

Because most people of that time were superstitious, they avoided onyx, a dark quartz mineral, because they feared it would bring hate, evil visions, strife, and bloodshed. To protect themselves from evil, people carried sardonyx, a quartz mineral that was believed to counteract the evil of onyx.

Even today, some people believe crystals have powers that cannot be explained by science. They acquire crystals, such as quartz or amethyst, to aid in healing, increase personal understanding, strengthen friendships, or cure loneliness.

The Giants

Although philosophers in ancient times were powerful, societies through the centuries contained doers as well as thinkers. These practical people worked with rocks and minerals to develop some of the first tools of science. Rather than relying upon myth and superstition to tell them how to work with gold or where to find a diamond, they used trial and error.

Philosophers looked down on these people who worked with their hands. But they are the "giants" on whose shoulders we stand to see into the future of crystals. Here are a few who made important discoveries while working with crystals:

16th Century: *Agricola*, a German physician, became a mining expert to improve his knowledge of medicinal minerals. He successfully classified many minerals according to color, weight, luster, taste, odor, shape, and texture. Today he is known as the Father of Mineralogy.

17th Century: *Sir Isaac Newton*, best known for his laws of motion, changed the direction of science with his theories and experiments on many subjects, including light. Quartz crystal glass, quartz prisms, and quartz lenses were key to his study of light. The telescope at Mount Palomar Observatory in California is a direct descendant of Newton's reflecting telescope.

18th Century: *René–Just Haüy*, a French scientist and priest, described the shape and angles of crystals by the mathematical arrangements of their particles. Haüy could not see these bits of matter, but his experiments with the electrical nature of crystals proved the existence of "particles" later identified as electrons. Scientists call him the Father of Crystallography.

With the Space Shuttle in the background, this artist's concept shows the powerful space telescope that scientists believe will show us the edge of the universe.

19th Century: *Friedrich Mohs*, an Austrian mineralogist, developed a scale of hardness in 1822 that is still used to help identify crystals. At the end of the chapter, you'll learn how to use this scale.

20th century: *Max von Laue*, a German physicist, beamed X rays (discovered by Marie Curie in 1898) through crystals onto photographic plates. His research in 1912 demonstrated how to make atoms visible and showed that atoms of crystals are arranged geometrically.

Physicists *Max Knoll* and *Ernst Ruska* discovered the electron microscope in 1929, improving magnification of specimens up to 12,000 times.

In 1985 researchers at Stanford University in California and at the IBM research laboratory in Zurich, Switzerland, simultaneously discovered the scanning tunneling microscope, which allowed scientists to see crystals form. That same year, *Albert Crewe* of the Enrico Fermi Institute in Chicago designed the scanning electron microscope. This powerful tool lets scientists see and control individual atoms as they lock into place within a manufactured crystal.

Try it Out

Mohs's Scale of Hardness

A simple scratch is used by scientists in a number of ways to determine the properties of crystals. Mohs's Scale of Hardness is one that uses a scratch to test the durability of crystals.

The Austrian mineralogist Mohs ranked minerals by their degrees of hardness with the numbers one through 10. Each mineral on the Mohs scale will scratch all specimens with a lower number. For example, feldspar, which has the number 6, will scratch slate (ranked as 3 on the Mohs scale) but it will not scratch topaz (an 8). A diamond has a hardness of 10; it can scratch all other natural substances.

Any crystal that scratches glass has the hardness of 5.5. Before you try this, make sure the glass you use is on a flat surface.

Most minerals represented on the Mohs scale can be purchased at rock or hobby shops, or you can use the tools listed here as substitutes for minerals.

Minerals	Tools
1 talc, graphite	
2 gypsum, galena	
	2.5 fingernail
3 calcite, slate	3 copper penny
4 fluorite, limonite	
5 apatite	
	5.5 knife blade, window glass
6 feldspar	

Minerals

7 quartz
8 topaz
9 corundum
  (ruby)
10 diamond

Tools
6.5 steel file

Gypsum is one mineral on Mohs's Scale of Hardness.

◆ FIVE ◆
How Nature Makes Crystals

Crystal balls cannot show us the future, but they hold secrets of the past. Within crystals we can find clues to the history of the earth and evidence of the geological forces that molded mountains and valleys.

Most of the 20,000 known crystallizing substances are manufactured. About 3,000 occur naturally as minerals in

Forces of nature, such as the Columbia Glacier in Alaska, wear away the earth's surface.

Magnetite, a form of igneous rock, from Magnet Cove, Arkansas

the earth's crust, but less than 50 of these are common components of rocks.

Regardless of where you are, you will be able to find quartz, feldspar, mica, and some form of limestone at your feet, or just below the earth's surface. All of these were formed by one of three major geologic processes. Understanding these processes will help you identify some crystals and unlock the secrets of their growth.

Igneous Rock

Six miles or more beneath the earth's surface is a molten layer of rock compacted by extraordinary pressure and heat. The weight of the earth above creates pressures of 60,000 pounds per square inch; the temperature is a blistering 1,600° F.

This molten solution of minerals, or magma, is constantly being squeezed upward into crevices of the earth's solid crust, creating pockets and veins of new material in older rocks.

Magma intrudes wherever there are weaknesses in the earth's crust, forming underground lakes or huge, vertical dikes that cut across older layers of rock. Sometimes the force of its intrusion creates earthquakes.

As molten magma rises toward the surface of the earth and cools into solid rock, crystals form. The speed of the cooling determines the size of crystals. Consider granite, which can be found in most sections of the country. Its crystals are mostly light-colored quartz, feldspars, and mica.

Cut beryl crystal with natural beryl from Colombia

If the crystals in a piece of granite are small, you know that the magma cooled quickly, probably because it was in contact with layers near the surface that halted crystal growth.

If the crystals are chunky, you can assume that they formed slowly, deep in the earth, and under good growing conditions. Such conditions allowed minerals to accumulate into a 36-ton crystal of beryl 60 feet long and 12 feet across. This record-breaker was discovered on the island of Madagascar, off the southeastern coast of Africa.

Closer to home, spodumene crystals more than 40 feet long have been found in the Black Hills of South Dakota. Though forged deep in the earth, solidified magmas eventually are exposed as the earth's surface gradually wears away.

Igneous rocks form from elements commonly found in magma. In order of abundance, they are oxygen, silicon, aluminum, iron, calcium, magnesium, sodium, and potassium. Each mineral crystallizes out of a molten solution in a specific order depending on the rate of cooling.

Red, green, and black minerals crystallize first at high temperatures. Many of these are heavy and metallic, such as magnetite (and, as its name implies, also magnetic).

The next minerals to crystallize include iron and manganese, both members of the pyroxene family. One familiar crystal of this group is green, fibrous jadeite. The amphiboles, another iron-manganese group, form next. Black hornblende crystals are typical of this group.

Lightweight, light-colored feldspars, which contain aluminum, crystallize next in colors of yellow, pink, green (microline), and white (moonstone). The last minerals to form crystals from magma are members of the quartz family.

The texture of igneous rocks is unique. Their grains fit together like jigsaw puzzle pieces because each set of minerals forms in the space left by those that crystallized before. Each mineral is firmly embedded in the surrounding particles and cannot easily be dislodged.

Not all melted rock crystallizes slowly beneath the earth's surface. Sometimes it shoots up to the sky when a volcano erupts. Lava, the melted rock that flows from a volcano, cools so quickly that only microscopic crystals form.

Obsidian is volcanic glass created when melted rock spews into the air

When lava shoots from volcanoes, such as Mount St. Helens in Washington State, the molten rock crystallizes instantly as it hits cool air.

and solidifies almost instantly. Crystals can even form from sulfurous gases that escape from volcanic vents.

About 100 miles below the surface of volcanoes, diamonds may form. Pushed to the surface in great cone-shaped pipes, many lie in the throats of old volcanic hills. The diamond mines of South Africa are of this type.

Sedimentary Rock

Wind, water, frost, and glaciers build other kinds of crystals as they constantly wear away the earth's surface. Loosened grain by grain over centuries, a mountain can be taken apart, strewn in layers across a valley floor, and consolidated by pressure into sedimentary rocks.

The texture of sedimentary rock usually is uneven.

Erosion reveals limestone deposits at Cedar Breaks National Monument in Zion National Park, Utah.

Individual grains stand out; they do not interlock like igneous crystals. Because many of them are formed in water, bits of organic matter, such as fossil remains, may be present.

Sedimentary rock takes many forms. Loose sand joins as sandstone. Layers of mud, clay, or silt are compacted into a finegrained shale, and seashells are pulverized into limestone.

Sedimentary rocks can be formed by chemicals as well as by pressure. Minerals dissolved in water crystallize as evaporation occurs. Enormous deposits of salt crystals form when desert lakes evaporate. Utah's Great Salt Lake is slowly drying up, and its water steadily becomes saltier.

Water seeping into underground deposits of limestone carves out caves and lines them with the crystal formations we call stalactites and stalagmites. Stalactites form when dripping water leaves deposits that make icicle shapes, and stalagmites build up from drops that fall to the ground.

Known as "The Pipe Organ," stalactites and stalagmites formed this attraction at the Howe Caverns near Cobleskill, New York.

Metamorphic Rock

Metamorphic means "changed form." It is the perfect name for this process, which can completely change the original features of igneous and sedimentary rocks.

As mountains rise, continents shift, and other earthshaking events occur, rocks are folded, squeezed, stretched, and crumpled. These tremendous pressures also generate intense heat. Both pressure and heat can change rock physically and transform minerals into ones that are more stable under the new conditions.

Crystals in this new rock usually are flattened and realigned into parallel bands found in gneiss (pronounced "nice") rocks. In gneiss, dark, thick bands of mica and hornblende alternate with light bands of quartz and feldspar.

Micaceous gneiss, a metamorphic rock, also may contain hornblende, quartz and feldspar.

In rocks called schists, flat mica crystals lie like so many stacks of paper. The weakness of mica allows one to split a schist along this layer. Crystals of quartz and other minerals may impart a salt-and-pepper appearance to schists, and many interesting minerals may be found in them: garnet, epidote, and hornblende, for example.

Quartzite is a shiny, compact rock that forms when sandstone grains are melted together under intense heat and cooled so quickly that their quartz crystals can barely be seen. Marble is another finely textured metamorphic stone wrought from the heating and compaction of limestone.

We have seen how magma changes when it meets cooler rocks, but imagine what contact with hot magma does to those rocks! The cooler rocks melt,

Vein of granite in mica schist found in New York State

mix with some of the magma, and form new metamorphic crystals.

When limestone mixes with magma and clay, crystals of garnet or corundum, a relative of ruby, may form. If sulfides are present, the result may be pyrite, or "fool's gold." Oxides may create magnetite, hematite, or spinel. Other combinations make crystals of amythest, garnet, topaz, and tourmaline.

Try it Out

Finding Crystals in Rocks

Scientists and rockhounds have developed many tests to identify and understand rocks, minerals, and crystals. Some of these are technical tools, but others can be used by

Hematite, like this specimen from Switzerland, is another form of metamorphic rock (top); an enormous metamorphic garnet weighing 9 ½ pounds

anyone. One simple method commonly used to identify minerals is a *streak test*.

Get a piece of unglazed porcelain, such as the back of a tile, and scratch the porcelain with an unidentified mineral. The color of the resulting streak is a clue to the mineral's identity. Here are some examples of what you may find:

Mineral	Streak Color
hematite	red
limonite	yellow
galena	dark gray
graphite	black
pyrite	black
fluorite	white
quartz	white

Here are some other methods used to unlock the secrets of crystals:
Color is easy to observe. Certain

Pyrite — known to prospectors in the Old West as fool's gold

minerals are known to occur in certain colors and to change color according to temperature.

Luster describes the way a mineral reflects or refracts light. Its texture can be described as glassy, waxy, or metallic.

Mohs's Scale of Hardness is a series of ten scratch tests for rating the hardness of a mineral. On the Mohs Scale, diamond is 10 and talc is 1. You'll find a Mohs Scale on page 53.

Cleavage describes the way a rock splits. This provides clues to the crystal structure of a mineral. There are five levels of cleavage from poor (the mineral bornite) to fair, good, perfect, and eminent (mica).

Fracture is another way to describe how minerals split, since not all break in perfect cleavage. Quartz, for example, breaks in a conchoidal (kong-KOI-dal) shape, like a

A *conchoidal fracture in volcanic obsidian*

hollowed out dish. Other fractures are called "uneven" and "earthy."

Specific gravity is a way of ranking minerals by their relative weight compared to an equal volume of water. For example, borax has a specific gravity of 1.7, while gold has a specific gravity of 19.3.

Electrical tests tell which minerals carry electrical charges.

Flame tests are used to note the color of a flame as a small amount of a mineral burns, or to determine the temperature at which a mineral will fuse in a blowpipe flame. *These tests are done in scientific laboratories and should not be attempted at home.*

Geiger counters detect minerals that have a radioactive charge.

Optical tests provide the most precise scientific ways to identify minerals in a rock or atoms in a crystal. For example, X rays show the arrangement of atoms in a mineral.

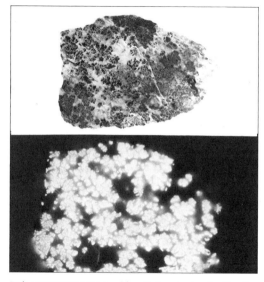

Radioactive uraninite in natural form (top), and captured on film after being left in darkness for a week

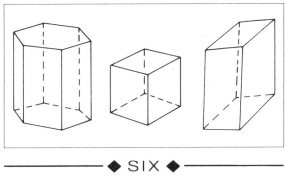

◆ SIX ◆
CRYSTAL FAMILY RESEMBLANCES

The incredible variety of crystals is dazzling. Where is the organization in this endless array of colors, shapes, and textures? Similarities between one mineral and another are not always easy to see. The best way to recognize resemblances among crystals is to learn about the "families" to which they belong.

Classic geometric shapes of crystal families

How can you tell which crystals are related? In human families, you might notice similarities in size, eye and hair color, skin tone, or facial characteristics. Within families of crystals, you'll look for different clues.

Let's consider quartz, diamonds, salt, and graphite. Are there any obvious similarities? You might count the sides, or faces, of each crystal; look at the way edges meet; or compare the size of the angle where two sides, or faces, meet.

If you paired quartz with graphite and diamonds with salt, you're a whiz. If you didn't, welcome to the club — nearly all of us must be shown the similarities before these family ties become crystal clear.

Each crystal has a predetermined building plan based on a specific arrangement of atoms, and the size of the angle formed where two faces join reflects how the atoms in a crystal have packed together.

Every crystal family has a basic shape, plus a number of variations on that form that all family members must fit. Gold, for example, may begin forming a typical cube-shaped crystal, but may bump into an obstacle that forces it to grow into a pyramid shape.

Six basic systems govern the way crystals grow, and there are 32 possible variations in shape.

Symmetry in Salt

Symmetry, or balance, is the key to the rules that govern crystal structure. To see an example of symmetry, let's start with salt — the simplest and most familiar of all crystals.

You will need:

paper (stiff paper or light cardboard works best)
scissors
glue
ruler
protractor
your crystal notebook

1. Trace the diagram for creating a paper model of a salt crystal on page 74.
2. Cut along the solid lines and fold on the dotted lines. Glue the tabs.

Now study the diagram on this page (figure 8) with three intersecting planes marked A, B, and C. Crystallographers call these planes "axes of symmetry." Imagine placing this figure inside your model of a salt crystal.

Think of each plane as a knife that is cutting the crystal in half. You will discover that

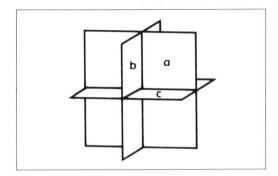

Figure 8: *Planes of symmetry in an isometric, or cubic, crystal*

no matter how it is divided along its axes, this model is symmetrical. Symmetry is the first rule of crystal growth.

This cube-shaped model of a salt crystal represents the family ideal of the *Isometric System*, or cubic system, to which gold, pyrite, diamond, magnetite, and garnet also belong.

Every crystal in this system is divided by three axes of equal length that make right angles (90 degrees) to one another.

The second factor that rules crystal growth is the length of each axis. If one axis of your model were a different length, how would that affect the shape?

The third factor that governs crystal growth is the size of the angles where crystal faces meet.

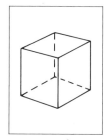

Isometric Family

Now that you have the tools, see how they help you understand the other five basic crystal families.

Tetragonal System. Zircon is a crystal gemstone from the tetragonal system.

The basic tetragonal shape is a rectangle with three axes that meet at right angles. The two horizontal axes are of equal length, but the vertical axis has a different length specific to each mineral.

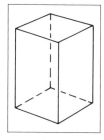

Tetragonal Family

Orthorhombic System. Topaz, olivine, sulfur, and barite belong to this family.

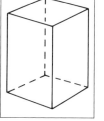

Orthorhombic Family

CHAPTER 6 71 ◆

Like the tetragonal system, the orthorhombic system also is rectangular, but its three axes are of unequal length. They intersect each other at right angles. The length of each axis is specific to each mineral.

Monoclinic System. Sugar is a monoclinic crystal, and so are some micas, talc, malachite, borax, gypsum, and azurite.

The basic monoclinic shape is a tilted rectangle with three axes of unequal length. Crystals in this family have two axes at right angles to one another, but the third axis makes an oblique angle — more than 90 degrees.

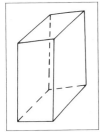

Monoclinic Family

Hexagonal System. Many gemstones occur in hexagonal forms, such as beryl, ruby, sapphire, tourmaline, quartz, and apatite.

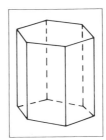

Hexagonal Family

There are four axes to the hexagonal system: three of equal length on a horizontal plane, 120 degrees apart, and a fourth axis that is vertical and which varies with each hexagonal mineral.

Triclinic System. Turquoise and polyhalite belong to this family.

Crystals in this form can be tilted, tabular, or triangular, but all have three axes of unequal length that form oblique angles.

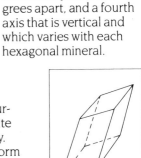

Triclinic Family

♦ 72 CRYSTALS

A crystal grown from a solution of monoammonium phosphate.

Try it Out

Classic Crystal Family Shapes

Use the following patterns to make paper models of the six basic crystal family shapes. Trace the pattern onto stiff paper or light cardboard. Cut along the solid lines, fold on the dotted lines, and then glue the tabs.

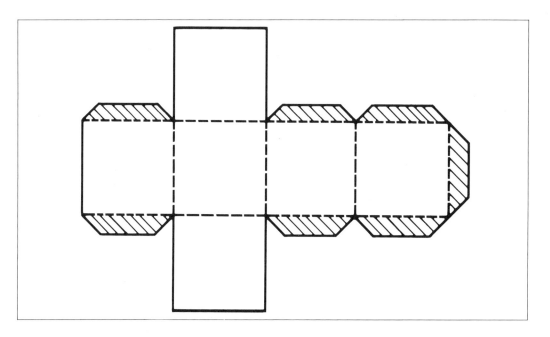

◆ 74 CRYSTALS

Isometric Crystal

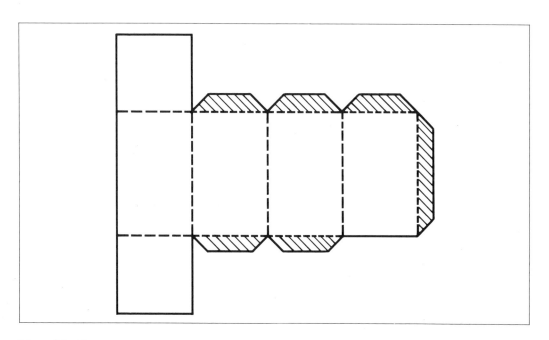

Tetragonal Crystal

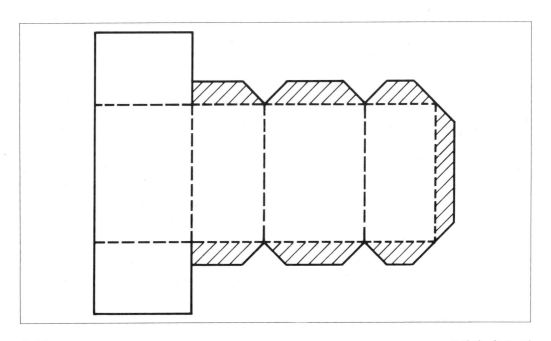

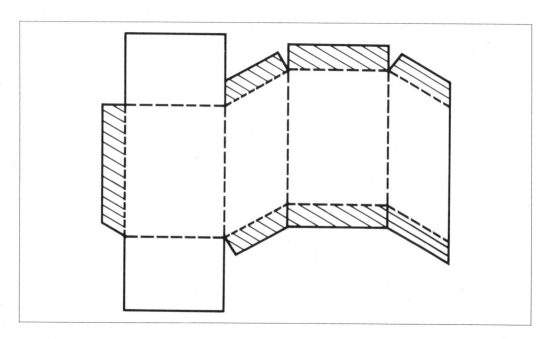

Monoclinic Crystal

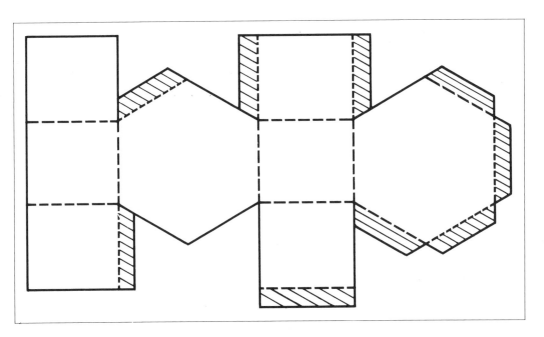

Hexagonal Crystal

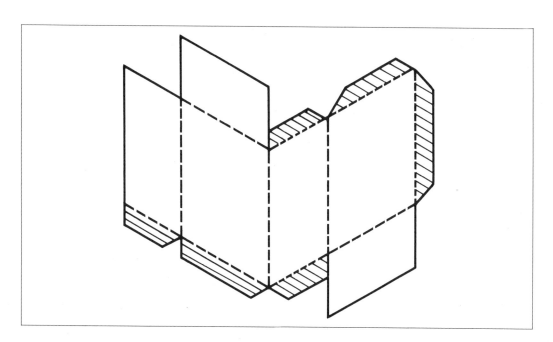

Triclinic Crystal

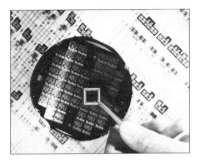

◆ SEVEN ◆
A New Age for Science

We are hurtling into a future filled with technology that will change our lives in ways we have not yet begun to visualize. New doors are opening, and they hinge on a working knowledge of subatomic behavior in crystals and on creativity in growing crystals to fit specific industrial needs.

The microchip in these tweezers can process as much information as a roomful of computers from the early 1970s.

The new technologies aim to bring us faster, more efficient communications, transportation, computers, medical tools, and energy sources.

The silicon chip that shuttles electrons in a millionth of a second is too slow. Now the goal is to reach the speed of light, to reduce a factory to the size of a human cell, to make something precious out of air, water, and sand.

Does this sound familiar? Are today's scientists modern alchemists? The goal is no longer one of turning common metals into gold; that has already been done in the laboratory simply by pushing the right atoms into the right crystal shape. Today, synthetic crystals can be grown atom by atom with the help of high-tech instruments that add or eliminate elements.

The Research Race

Scientists are competing to find the fastest, least expensive way to move energy. These researchers are in a relay race to develop new technology, and the strides that one team makes could help another cross the finish line first. The "runners" in this race are:

Silicon. Some scientists are continuing to refine silicon's shape, believing its limitations as a semiconductor can be designed away. For example, scientists have figured out a way to grow microscopic circuitry directly into a silicon chip, rather than etching it in with a laser.

Superconductors. Several laboratories are using crystals to develop superconductors that transport electricity faster than ever before. But there are drawbacks at the present; superconducting crystals function only

Magnified quasicrystals, a new form of solid state matter, were discovered in 1985 by an Israeli scientist working at the U.S. National Bureau of Standards. Quasicrystals may be useful for a variety of things, from airplane parts to high-quality resistors.

in a super-cooled nitrogen or helium environment, and many of them are rare or poisonous.

Optoelectronics. Some expect photon emission to be the energy wave of the future. Crystals of a manufactured synthetic called gallium arsenide emit light when stimulated, absorb sunlight more efficiently than silicon, resist radiation, function at higher temperatures, and conduct electricity three to six times faster than silicon.

Instead of using metal circuitry like that added to silicon chips, gallium arsenide chips use fiber optics — packets of light racing through tiny glass tubes. Plans are on the drawing board for advanced computers and solar cells that will work because of these, or other types, of crystals.

Biochips. A few investigators stake their claim on the biochip. These scientists are working to transform protein molecules into tiny computers. The advantage to biochips would be one of size — a protein molecule is much larger than a silicon molecule, and holds even more potential for computing tasks.

As all these technologies are being developed, new uses for crystals are reported nearly every day. One of these is a silicon sensor — no larger than a pencil eraser — that can detect the presence of hydrogen or other potentially dangerous gases. Hydrogen is odorless and colorless, but if it collects undetected in a laboratory, it can explode.

More of these microsensors are being developed to detect harmful substances in foods, such as pesticide residues on fruits and vegetables. The technology is based on piezoelectric crystals, using proteins that react with specific chemical compounds to generate changes in the rate of vibrations within the crystals.

Some researchers are working on a medical monitor contained in a silicon-coated "pill." Here's how scientists think it will work: You would swallow the pill, and tiny instruments inside it would analyze information on body functions — your heart rate or gastric pressure, for example — and then the pill would beep this information to a receiver sewn into your shirt.

Just as exciting as the new and future applications of crystals is the technology of producing them. Spacelab 1 and 2 offered an opportunity to grow larger, more perfect protein crystals.

In space, atoms move freely around the

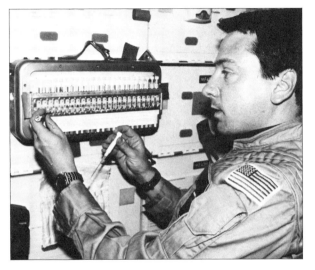

Aboard the Space Shuttle, payload specialist Charles D. Walker conducts experiments to crystallize proteins.

crystal seed and attach themselves according to the logic of the seed crystal's pattern. On earth, gravity pulls atoms down, causing crystals to become malformed. The weightlessness of space allows crystals to grow evenly on all sides.

These perfect, completely pure crystals could become superconductors that function at normal temperatures on earth, and they could be grown from materials already available in space.

Nanotechnology: A New World

Futurists believe that technology ultimately will be microscopic. They envision a world in which food or clothing would be manufactured atom by atom, probably by protein "assemblers."

This new field, known as nanotechnology, encompasses more than crystals, but crystals will play an important role. Nanotechnological engineers now use crystal models to think about how protein assemblers could work. These researchers foresee proteins imitating the work of living cells while building matter from such plentiful sources as air, water, sun, and sand.

How will nanotechnology work? Scientists don't yet know, although they are dreaming of fantastic ways to apply their theories.

Futurists believe vending machines won't just store food — they'll create it. Medical technicians envision tiny computers tucked into the corner of a human cell that would sense damage within the body and repair it.

Emerald green with envy — that might well describe the jealousy of a medieval alchemist who suddenly materialized today. The search for the Philosopher's Stone was just a step in the long journey toward understanding crystals and developing technology in a new age of science that promises improved communications, faster transportation, and abundant energy sources.

GLOSSARY

CRYSTAL. Solid matter which has a symmetrical pattern of faces and angles caused by an orderly, repetitive arrangement of atoms.

CRYSTALLINE. A substance whose molecules are arranged like those of crystals, but which has no defined faces and angles.

CRYSTALLOGRAPHY. The science of crystals and crystal growth.

IGNEOUS. A rock-forming process that produces crystals as superheated molten rock is cooled deep in the earth.

METAMORPHIC. A rock-forming process that creates crystals when geologic forces change igneous or sedimentary rocks.

MINERALOGY. The science and classification of minerals.

ORGANIC. Matter that is living or that was once alive. Organic materials, such as plants, are not crystals, but they may contain chemicals in crystal form.

SEDIMENTARY. A rock-forming process in which wind, water, frost, or glaciers wear away the earth's surface and deposit layers of sand, clay, or organic material that consolidate into rock.

SEMICONDUCTOR. A silicon chip, etched with electrical circuits, used to power computers and other electronic equipment.

SILICON. The most abundant element on earth, next to oxygen. Because its atoms are arranged in a specific pattern, silicon is a preferred base for semiconductors.

INDEX

You'll find additional information in the Glossary, p. 86.

Agricola, 51
Alchemists, 29, 49
Alum, 4, 11
Alunite, 4
Amethyst, 50, 64
Amphiboles, 58
Apatite, 72
Aristotle, 49
Arrowheads, 25
Atomic theory, 49
Axis length, 71
Azurite, 39, 72

Bacon, Roger, 30–31
  light experiments of, 34–35
Baking soda, 11
Barite, 71
Beryl, 58, 72
Biochips, 83–84

Bismuth, 33
Blue Tavernier diamond, 40–41
Borax, 72
Boric acid, 12
Bornite, cleavage of, 66
Brownstone, 15
Building materials, 15

Carnelian, 39
Chert, 26
Clay, 61
Cleavage, 66
Concrete, 15
Corundum, 64
Crewe, Albert, 52–53
Crystal. *See also* Quartz
  balls, 28
  goblets, 13
  notebook, 19

Crystallography, Father of, 51
Crystals. *See also individual names of crystals*
  beliefs in curative powers of, 50
  characteristics of, 9–10
  definition of, 8, 9–11
  families of, 68–73
  field guide to, 17–19
  in foods, 11, 12
  formation of in nature, 55–67
  found indoors, 11–14
  found outdoors, 14–17
  freeze-dried, 12
  growing own, 4–6, 20–23
  inspection of, 18–19
  movement of, 32–33
  poisonous, 18
  precious, 36–46
  protein, 83–84

in space, 83–84
structure of, 69–73
subatomic behaviors of, 80
superconducting, 81–82
synthetic, 81, 82
Cullinan diamond, 41
Curved lenses, 29

Democritus, 49
Diamond coating, 14
Diamond mines, 60
Diamond(s), 36, 37, 39–41
composition of, 42–46
hardness of, 66
in rough, 41–42
shape of, 69, 70
Dry mixes, 12

Electrical tests, 67
Electricity, 81–82
Electron microscope, 52
Electronics, 33

Electrons, 51
Emeralds, 37
Epidote, 63
Epsom salts, 12
Evaporation, 5–6
Eyeglasses, 30. *See also* Lenses

Feldspar, 15, 57–59
in gneiss, 62
Fiber optics, 28, 82
Flagstone, 15
Flame tests, 67
Flea glasses, 30
Flint tools, 26
Fool's gold, 64
Fracture, 66–67
Freeze-dried crystals, 12
Freezer frost, 12, 17
Fructose, 11
Futurists, 84–85

Galileo, 31

Gallium, 33
Gallium arsenide crystals, 82
Garnet, 63, 64
shape of, 70
Geiger counters, 67
Gemstones, 28–29, 36–46
Geologic processes, 56–64
Glass, quartz, 28–31
Glassmaking, 28
Gneiss, 62
Gold, 36
in alchemy, 49
crystal shapes of, 70
Granite, 15, 57, 58
Graphite crystals, 14
composition of, 45–46
structure of, 69
Great Salt Lake, 61
Greek philosophy, 48–49
Gypsum, 72

Hardness scale, 52, 53, 66

Haüy, René-Just, 51
Helium, supercooled, 82
Hematite, 15, 64
Hexagonal system, 72
Hexahedron shape, 46
Hope, Henry Thomas, 41
Hope diamond, 40–41
Hornblende, 58, 62, 63
Hydrogen, 83

Igneous rock, 56–60
Ilmenite, 15
Iron, 58
Iron-manganese crystals, 58
Isometric system, 70–71

Jade, 37-39
   characteristics of, 10
Jadeite, 58
Jasper, 26
Jewelry, 13
   gemstones in, 39

Kitchen crystals, 11–12
Knoll, Max, 52
Koh-i-noor diamond, 39

Lapis lazuli, 39
Laue, Max von, 52
Lava, 59
LED displays, 14
Lenses, 29–30
Light, nature of, 31
Limestone, 61, 64
   compaction of, 63
Lithium, 33
Luster, 66

Magma, 57, 58, 59
   effects of, 63–64
Magnetite, 15, 58, 64
   shape of, 70
Magnifiying lens, 29
Malachite, 39, 72
Manganese, 58

Marble, 63
McLean family, 41
Medical monitors, 83
Medical technology, 85
Metals, 12–13. *See also individual names of metals*
   precious, 36
   youth of, 49
Metamorphic rock, 62–64
Mica, 15, 57, 62, 63, 72
Microline, 58
Microscopic circuitry, 81
Microsensors, 83
Middle Ages, mysteries of, 49–50
Minerals. *See also individual names of minerals*
   color of, 65–66
   crystallization of, 58, 59
   fracture of, 66–67
   luster of, 66
Mineralogy, Father of, 51
Mohs, Friedrich, 52, 53

Molten rock, 56–57
Monoclinic system, 72
Moonstone, 59
Mountain of Light diamond, 39
Mud, 61

Nanotechnology, 84–85
Nero, 29
Newton, Sir Isaac, 31, 47–48, 51
Nitrogen, supercooled, 82
Noble metals, 49

Obsidian, 15, 59–60
Olivine, 15, 71
Omens, 50
Onyx, superstitions about, 50
Opal, 36
   superstitions about, 50
Optical tests, 67
Optoelectronics, 82–83
Organic materials, 19
Orthorhombic system, 71–72

Oxides, 64

Pencil lead, 14
Philosopher's Stone, 49
Photocommunications, 28
Piezoelectric crystals, 83
Plant food, 14
Plastics, 19
Polyhalite, 72
Polymers, 19
Potassium aluminum sulfate. *See* Alum
Prisms, quartz, 31
Protein assemblers, 85
Protein
   crystals in space, 83–84
   molecules, 83
Pyrite, 64
   shape of, 70
Pyroxene crystals, 58

Quartz, 24–25, 57
   as building material, 15
   characteristics of, 10
   in electronics, 33
   fracture of, 66–67
   in glasses, 13, 28–31
   in gneiss, 62
   in healing, 50
   history of, 28–31
   making rainbows with, 34–35
   mineral crystals of, 59
   movement of, 32–33
   myths about, 49
   Newton and, 51
   shape of, 72
   in schists, 63
   silicon revolution and, 26–27
   in Stone Age inventions, 25–26
Quartzite, 63

Radio waves, 33
Research race, 81–84

Rock
  igneous, 56–60
  metamorphic, 62–64
  sedimentary, 60–61
Rock salt, 11
Rubies, 36, 37, 72
Ruby bearings, 13
Ruska, Ernst, 52

Saccharin, 11
Salt crystals, 11
  building, 20–23
  deposits of, 61
  symmetry in, 69–72
Sand, 15, 61
  beach, 15
  in construction, 15
  desert, 16
Sandpaper, 14
Sandstone, 61
  melted, 63
Sapphires, 36, 37, 72
Sardonyx, 50
Scale of Hardness, Mohs's, 52, 53, 66
Scanning tunneling microscope, 52–53
Schists, 63
Scientific discoveries, 47–53
Sea salt, 11
Seashells, pulverized, 61
Sedimentary rock, 60–61
Semiconductors, 33, 81
Shale, 61
Silicon, 82
  heat limitations of, 33
  research in, 81
Silicon chips, 26–27, 33, 81
Silicon dioxide. *See* Quartz
Silicon sensor, 83
Silicon-coated "pill" monitor, 83
Silt, 61
Silver, 36
Snowflakes, 16–17
Soft drink mixes, 12
Solar cells, 82–83
Space, crystals in, 83–84
Spacelabs, 83
Specific gravity, 67
Spectacles, 29–30
Spinel, 64
Spodumene crystals, 58
Stalactites, 61
Stalagmites, 61
Star of India, 41
Stone Age tools, 25–26
Stones, clues in, 50
Streak test, 65
Subatomic behavior, 80
Sugar crystals, 11
  shape of, 72
Sugar substitutes, 11
Sulfides, 64
Sulfur, 71
Sulfurous gas crystals, 60
Super salt crystals, 20–23

Superconductors, 81–82
Supercooling, 4–5
Supersaturated solution (SSS), 21
Superstitions, 50
Surgery, 28
Symmetry, 69–73
Synthetic crystals, 81, 82

Talc, 72
Technologies, new, 80–85
Telecommunications, 28
Telescopes, 30–31, 51
Tetragonal system, 71
Tetrahedron shape, 44–45
Topaz, 64, 71
Tourmaline, 64, 72
Triclinic system, 72
Turquoise, 72

Volcanic eruption, 59
Volcanoes, 59–60

Watch crystals, 13
Window frost, 17
Winston, Harry, 41

X rays, 52, 67

Yttrium, 33

Zircon, 71

ABOUT THE AUTHOR

Marlene M. Robinson is the author of three science discovery books for children: *Who Knows This Nose?*, *What Good Is A Tail?*, and *Who's Looking At You?* She has taught science at elementary and junior high schools in New York City, and she is an education specialist who has developed interpretative exhibits and family programs for zoos in Boston, Philadelphia, and Washington, D.C.